HOW CAN I EXPERIMENT WITH ... ?

ELECTRICITY

Cindy Devine Dalton

Cindy Devine Dalton graduated from Ball State University, Indiana, with a Bachelor of Science degree in Health Science. For several years she taught medical science in grades 9-12.

Teresa and Ed Sikora

Teresa Sikora teaches 4th grade math and science. She graduated with a Bachelor of Science in Elementary Education and recently attained National Certification for Middle Childhood Generalist. She is married with two children. Ed Sikora is an Aerospace Engineer, working on the Space Shuttle Main Engines. He earned a Bachelors of Science degree in Aerospace Engineering from the University of Florida and a Masters Degree in Computer Science from the Florida Institute of Technology.

rourkeeducationalmedia.com

www.rourkeeducationalmedia.com

PROJECT EDITORS
Teresa and Ed Sikora

PHOTO CREDITS
Gibbons Photography
PhotoDisc
Walt Burkett, Photographer

ILLUSTRATIONS
Kathleen Carreiro

EDITORIAL SERVICES
Pamela Schroeder

Library of Congress Cataloging-in-Publication Data

Dalton, Cindy Devine, 1964-
Electricity / Cindy Devine Dalton
p. cm. —(How can I experiment with?)
Includes index
ISBN 978-1-58952-011-0 (Hard Cover)
ISBN 978-1-58952-016-5 (Soft Cover)
1. Electricity—Juvenile literature 2. Electricity—Experiments—Juvenile literature
I.Title

QC527.2 D35 2001

Printed in China, FOFO I - Production Company
Shenzhen, Guangdong Province

rourkeeducationalmedia.com
customerservice@rourkeeducationalmedia.com • PO Box 643328 Vero Beach, Florida 32964

Electricity: A form of energy that is made up of charged particles.

Quote:

"Discovery consists of seeing what everybody has seen and thinking what nobody has thought"

-Albert Szent-Gyorgyi

Table of Contents

What Is Electricity? 7
Why Does Matter Really Matter? 8
Static Electricity 10
Opposites Attract 12
What Is a Conductor? 14
What Is an Electric Current and Where
Does It Come From? 17
What Is an Electric Circuit? 18
An Electric Circuit With a Switch 20
Electromagnetism 22
Hands on: Swinging Cereal 24
Safety Dos and Don'ts 27
What Do You Know About Electricity? ... 28
Glossary 30
Further Reading 31
Websites to Visit 31
Index 32

What Is Electricity?

Electricity works in many ways. We usually think of electricity as the "thing" that provides the energy we use in our houses. Right? That is true. However, there is a lot more to electricity than that. It is a form of energy that runs machinery. It can be changed into other types of energy such as heat, light, and motion.

Big power plants such as this provide electricity for our homes and businesses.

Why Does Matter Really Matter?

Everything you see, hear, taste, and smell is made of **matter**. Matter is made of very small parts called **atoms**. Atoms are made of **protons**, **electrons**, and **neutrons**. Electricity is transferred or moves through the flow of electrons.

Everything you see, hear, taste, and smell is made of matter.

Static Electricity

Have you ever rubbed your socks on the carpet and then touched someone? Did you see a spark? You created static electricity. Static electricity builds when a solid, like your socks, is forced to free electrons. Your socks lose electrons when you rub them on the carpet, leaving a "hole." Electrons have a negative charge. Because you take electrons away, the "hole" has a positive charge. You have created an electric field. Both positive and negative charges want to move through the electric field. When you touch someone, you start a current known as static electricity. That's the spark you see. A current of static electricity only lasts a few seconds. We can't use it for energy in our homes.

It is a lot of fun to create static electricity. It is a very short burst of electricity!

Opposites Attract

Another way to see static electricity is to rub a balloon on your head. Then pull the balloon off and—wow—you've created static electricity! What you've done is rub off some of the negatively charged electrons from your hair. The balloon has the extra negatively charged electrons. It wants to attract, or pull, toward your hair, which has a positive charge.

Remember the protons we talked about earlier? They have a positive charge. They are out of balance without the electrons' negative charge. Have you ever heard the saying "opposites attract"? It's true! Negatively charged electrons are attracted to positively charged protons.

Static electricity can give you a bad hair day!

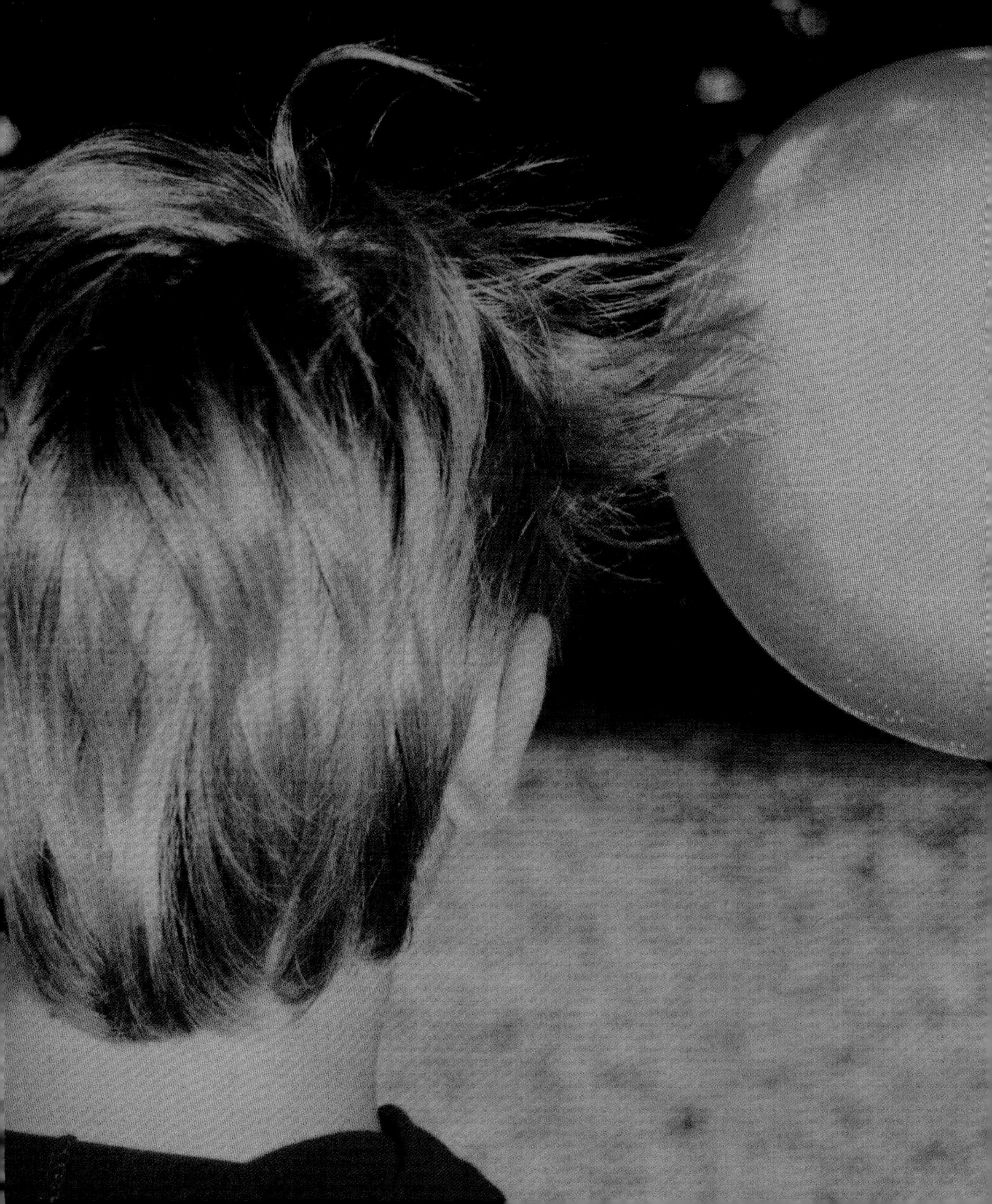

What Is a Conductor?

A **conductor** is something that allows electrons and electricity to flow. Most metals are good conductors. That is why you see silver and copper wires in lamps, stereos, and telephone poles. Poor conductors do not allow electrons or electricity to flow through them. Examples are rubber, plastic, or glass. These are good **insulators**. They keep electrical charges from escaping.

The metal wires on electrical appliances are inside plastic or rubber to make them safer to use.

What Is an Electric Current and Where Does It Come From?

You know that lightning is electricity. In lightning electrons move quickly and then stop. An electric current is different—the electrons keep moving through a conductor, like a wire.

Power plants make electricity and send it through a cable to your house. There are different sources of current electricity including chemical reactions from a battery, or the burning of coal or oil at power plants. Solar, water, and nuclear power can also be turned into electricity.

Batteries create an electric current to power this game.

What Is an Electric Circuit?

An electric **circuit** is a path along which electrons can move without being stopped. Look at the picture. A circuit has a source of electrons, like a battery, that produces an electric charge. It also has a conductor, like a wire, that connects to the light bulb and back to the battery. This is a complete circuit. The electrons move from the battery along the conductor to the bulb. The bulb lights. Then the electrons move back to the battery.

Everything that runs using electricity has an electric circuit.

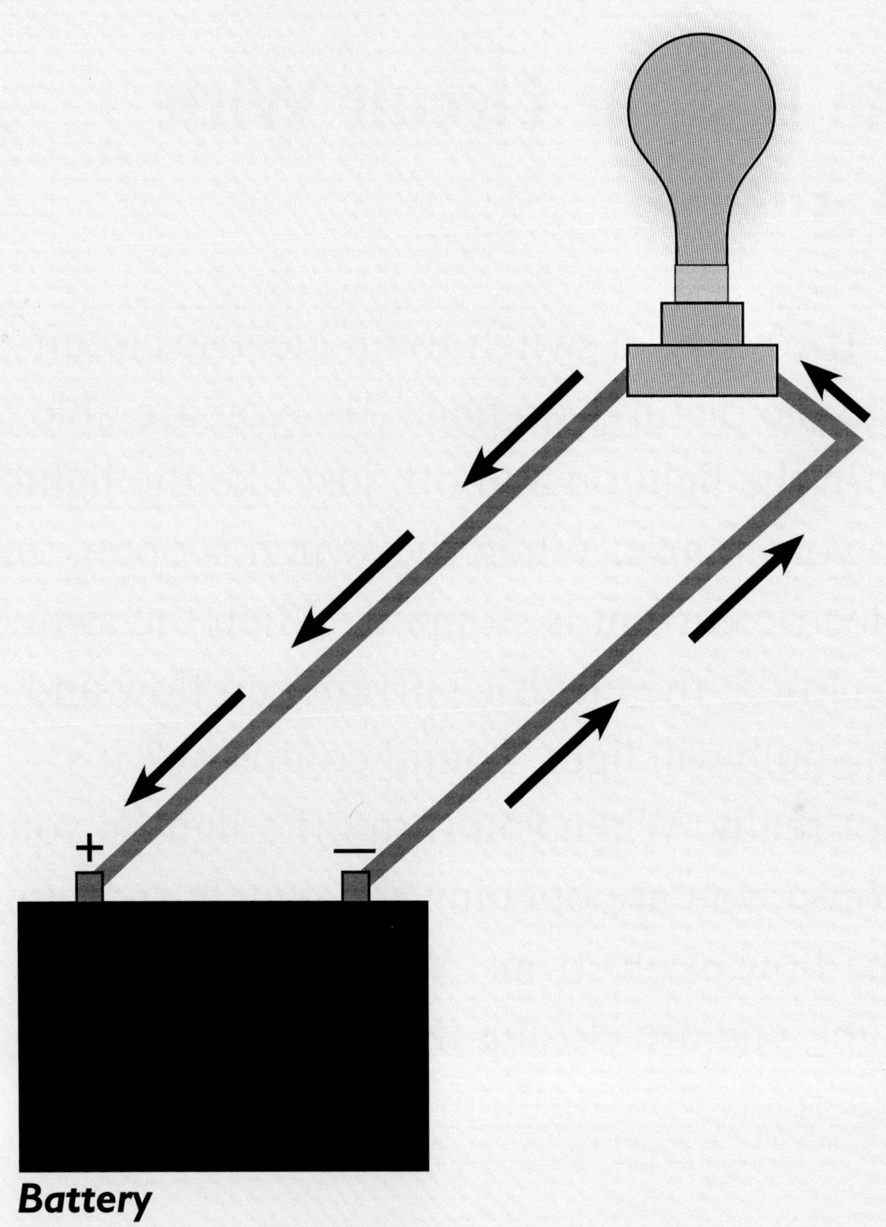
+
−
Battery

An Electric Circuit With a Switch

If you add a switch to an electric circuit—see the pictures on page 21—you are able to turn the light on and off, just like the lights in your house. When the switch is open, the electric current is stopped. When the switch is closed, the electric current can flow and the bulb will light. Sound confusing? It's not really. When you turn off a light in your house, you are opening the switch, stopping the flow of electrons. When you turn on a light, you are closing the switch, letting the electrons flow through the circuit.

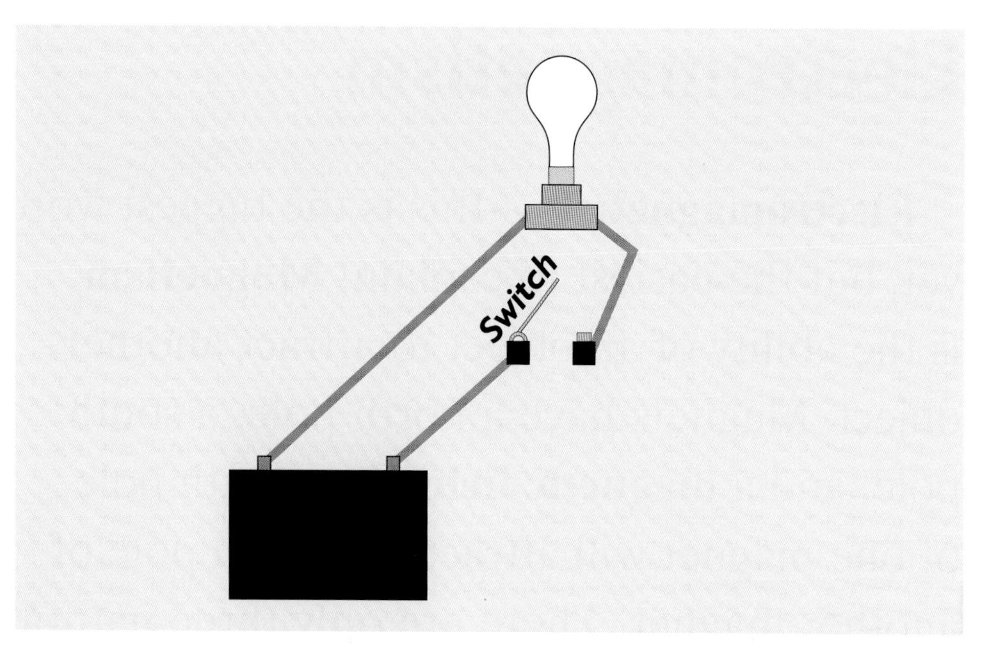

When the switch is open, the current is stopped and the light bulb stays off.

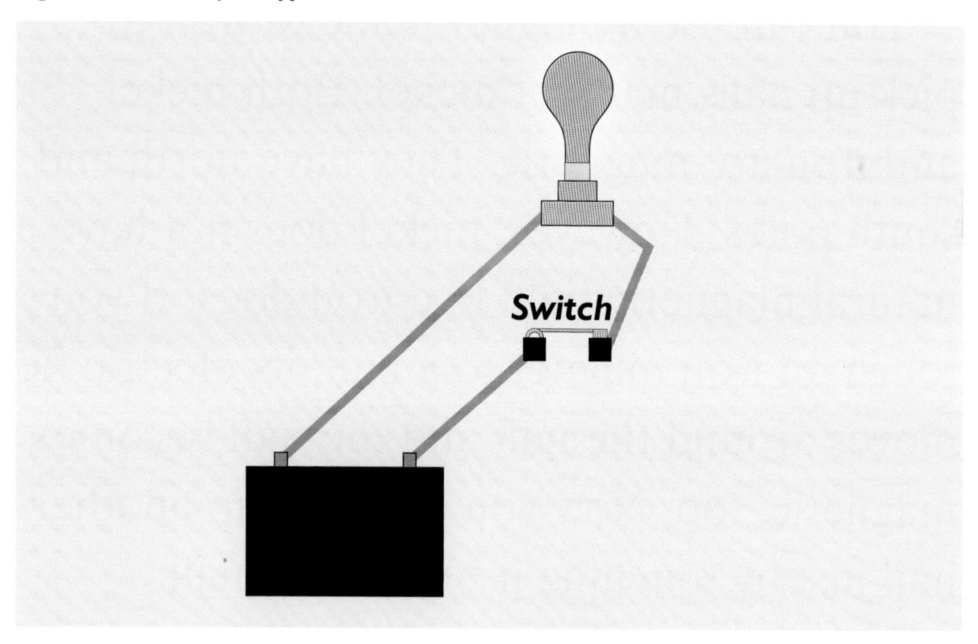

When the switch is closed, the current can flow and the light bulb comes on.

Electromagnetism

Electromagnetism—this is the biggest word yet, but the easiest to explain! **Magnetism** is the ability of an object to attract another object. Magnets have a north pole, a south pole, and a **magnetic field**. The south pole of one magnet will attract the north pole of another magnet. There are only three natural magnets on earth—iron, cobalt, and nickel.

You can use a magnet made of iron to pick up nails made of nickel. Both nickel and iron are magnetic. They have north and south poles. However, nickel does not have a natural magnetic field like iron does. If you touch an iron magnet to a nail, its electric field moves around the nail and the nail becomes magnetic, too. Now you can pick up another nail just by touching it to the first nail.

A compass uses the magnetic fields of the Earth and the north and south poles to point you in the right direction.

SPORTIN

Hands on:

Swinging Cereal

What you need:

- A balloon
- Thread
- Small pieces of dry cereal (O-shaped or puffed wheat)
- A wool sweater or blanket

Try This:

1. Tie a piece of the cereal to one end of a 12-inch piece of thread.
2. Find a place to tie or tape the other end so that the cereal does not hang close to anything else. (You may tape the thread to the edge of a table, but check with your parents first.)
3. Wash the balloon to remove any oils and dry it well.

4. Rub the balloon on a wool sweater to add a charge.
5. Slowly bring the balloon near the cereal. The cereal will swing to touch the balloon. Hold it still until the cereal jumps away by itself.
6. Now try to touch the balloon to the cereal again. The cereal will move away as you move the balloon closer.

The cereal is attracted quickly to the balloon.

What Happened?

Rubbing the balloon on the wool moved the electrons from the wool onto the balloon. The balloon had a negative charge. The cereal had no charge and was attracted to the negative balloon. When they touched, electrons moved from the balloon to the cereal. When they both had the same negative charge, they moved away from each other.

Safety DOs and DON'Ts

! Don't touch anything that is electrical when your hands are wet.

! Never play with light switches, electrical sockets, cords, or appliances.

! Always pull a plug by the plug, not the cord.

! Never stick anything but an electrical plug into an electric outlet.

! Don't use your radio or blow dryer near the bathtub or sink.

! Never touch anything electrical if you are standing in water.

! If your electrical cord is frayed, tell an adult right away.

! If your plug or outlet sparks, tell an adult right away.

! Turn all electrical appliances off during a storm, and stay out of the shower and off the phone. Lightning travels through cords and pipes!

What Do You Know About Electricity?

(Answers are at the bottom of page 29)

1. True or false?
   A conductor is something that allows electrons and electricity to flow.

2. What charge does an electron have?
   __ Positive
   __ Negative
   __ Neutral
   __ A slow one

3. What is one of the best materials for conducting electricity?
   __ Metals
   __ Wood
   __ Chalkboards
   __ Water

4. Everything you can see and touch is made up of:

__ air
__ solids
__ gases
__ matter

5. How much cleaner are electric cars than gasoline-fueled cars?

__ 100%
__ 50%
__ 0%
__ 97%

6. If you see the words "Danger: High Voltage" on electrical equipment, you should:

__ stay far away from the equipment and play somewhere else
__ take a picture of it
__ set up a lemonade stand next to it

1. True 2. Negative 3. Metals 4. Matter 5. 97%

6. Stay far away from the equipment and play somewhere else

Glossary

atoms (AT emz) — tiny particles made up of protons, neutrons, and electrons

conductor (ken DUK ter) — a material that can pass on electricity

circuit (SUR kit) — the path electricity passes along

electromagnetism (ih lek troh MAG neh tiz em)— magnetism made by a current of electricity

electrons (ih LEK tronz) — parts of an atom that have a negative charge

insulators (IN se layt erz) — materials that keep electricity from passing through

magnetism (MAG neh tiz em) — attracting metals, or producing a magnetic field, as with a magnet

magnetic field (mag NET tik FEELD) — the area near a magnet.

matter (MAT er) — what everything is made of

neutrons (NOO tronz) — the parts of an atom that have no charge

protons (PROH tonz) — the parts of an atom that are inside the nucleus and have a positive charge

Further reading

Exploring Physical Science, Prentice Hall, 1999
Let's Wonder About Science, Rourke Press, 1995
Energy and Action, Rourke Press, 1995

Websites to Visit

www.studyweb.com
www.askanexpert.com
www.howthingswork.com

Index

circuits 18, 20
conductor 14, 17, 18
electrons 8, 10, 12, 17, 18, 20, 26
electromagnetism 22
insulators 14
magnetism 22
matter 8
safety 27
static electricity 10